First published in 2005 by Prufrock Press Inc.

Published in 2021 by Routledge
605 Third Avenue, New York, NY 10017
2 Park Square, Milton Park, Abingdon, Oxon OX14 4RN

Routledge is an imprint of the Taylor & Francis Group, an informa business

Copyright © 2005 by Taylor & Francis Group

ISBN: 9781593630898 (pbk)

DOI: 10.4324/9781003236283

Contents

Contents

Information for the Instructor

Pack your gear and get ready to take your students on a safari. But on this safari, you won't be stalking wild animals. You'll be stalking clues that you can piece together to solve deductive logic puzzles. The reward for this expedition will be a bounty of thinking skills. *Logic Safari* gives students an opportunity to sort through related bits of information by combining, relating, ordering, and eliminating. The result is the logical linking together of ideas that leads to the puzzle's solution.

Each puzzle in *Logic Safari* has three parts. These parts are:
1. **The introduction** - This paragraph sets the background and helps students become familiar with the elements of the puzzle.
2. **The clues** - The clues relate all of the components and provide a basis for the logical linking together of the pieces of information, thereby allowing students to make deductions that will lead to the solution.
3. **The grid** - The grid provides a worksheet for sorting, eliminating and associating the clues. Every square on the grid represents a possible answer. By eliminating possibilities, one is finally left with only one choice per row or column. The one square that is not been eliminated is one correct solution. When this is done for every row and column, the puzzle solution is complete.

Any marking system for the grid is valid if it is used consistently. Many students prefer to use an **X** in a square to represent elimination of a choice and an **O** to represent a correct answer. Using "yes" and "no" works equally well.

In addition to these three parts, students may wish to jot down notes on scratch paper. This may help them in putting the information in rank order or in visualizing the relationships. It should be stressed that there is always more than one way to correctly solve a puzzle; and in sharing the way in which each person used the clues to arrive at the solution, students will gain insights into different modes of thinking.

Students will find that it is necessary to look not only at each clue individually, but also to look at the clues in relation to one another in order to derive as much information as possible. For example, if the clues state, "Gina is older than Eric and the girl with the teddy bear but younger than Gonzales," we can deduce a lot of information by the proper arrangement of the clues. If there four people in the puzzle, we know that Gina is second in rank of age. If there are two boys and two girls, we also know that Gonzales is a boy. We can also deduce that Mary is not Gonzales, Eric is not Gonzales, Gonzales does not have a teddy bear, Gina does not have a teddy bear, and Eric does not have a teddy bear. In addition, we know that Gonzales is the oldest. With one clue, then we have been able to make several eliminations and two positive connections.

There are three books in this series, so students are able to move from easy to intermediate levels of difficulty in deductive thinking. *Logic Safari* puzzles are an excellent way to strengthen students' logical deductive thinking skills. Students find the puzzles very motivating, and as they work with these puzzles they grow in their abilities to sort through information and make connections.

Birthday Bashes

Lyle, Ryan, and Troy have their birthdays in the same month. They have chosen to have their parties at the roller rink, the miniature golf course, and the pizza parlor. Tear open these clues to find out how each person will be celebrating his special day.

Clues

1. Lyle, and the boy who has chosen the roller rink, and the boy whose party will be at the pizza parlor are all very excited.

2. Troy's party will not be at the pizza parlor.

	roller rink	miniature golf	pizza parlor
Lyle			
Ryan			
Troy			

Swimming Lessons

Steve, Andy and Ted all go to the pool for swim lessons. They have different instructors. Today one boy is in a group learning to tread water, one is in a group diving for pennies on the bottom of the pool, and one is in a group jumping off the diving board. Dive into these clues to find out who is doing what.

Clues

1. Andy, his brother who is treading water, and his friend who is diving for pennies all live within walking distance of the pool.

2. Steve and Andy are not brothers.

	treading	pennies	diving board
Steve			
Andy			
Ted			

After School

Daralynn, Kevin, and Tara are three friends who stay for the after-school play program. Today one is playing with a yo-yo, one with a jump rope, and one with puzzles. Unlock the clues to find out who is doing what.

Clues

1. Daralynn, the boy with the yo-yo, and the girl with the jump rope are all picked up by their parents around 5:30.

2. Tara and Daralynn are sisters.

	yo-yo	jump rope	puzzles
Daralynn			
Kevin			
Tara			

Spring Musical

The second grade is working very hard on its spring musical. Megan, Adam, and Jeff have special roles in the production as a baseball player, a rose bud, and a young fisherman. Use these clues to find out who had what role.

Clues

1. Megan, and the baseball player, and the fisherman have to bring their own costumes and props.

2. Jeff has to remember to bring his bat and mitt to rehearsals as well as to team practice.

Aluminum Can Drive

Carla, Patrick, and Justin all wanted their class to be the leader in the aluminum can drive. They brought in 2, 3, and 5 bags of crushed cans. Smash your way through this clue to find out who brought how many bags.

Clue

1. Carla brought more bags than Justin but fewer than Patrick.

	2 bags	3 bags	5 bags
Carla			
Patrick			
Justin			

Helping in Grandpa's Garden

Tyler, Jack, and Brenda are helping their grandfather in the garden. One is transplanting seedlings from the hot bed, one is working compost into the soil, and one is tying tomato plants to stakes. Cultivate these clues to find out who is doing what.

Clues

1. Jack and the boy transplanting seedlings are happy when they get a chance to help their grandfather.

2. Jack and the girl staking tomatoes are cousins.

Earth Day

Children helping with the Earth Day celebration were encouraged to wear a T-shirt depicting something about the earth or nature. Janis, Charlie, Nicole, and Ashley wore shirts with hummingbirds, rabbits, the world, and the Rocky Mountains on them. Explore these clues to find out who wore each shirt.

Clues

1. Charlie, the girl who wore the hummingbirds, the girl with the rabbits, and the girl whose shirt portrayed the Rocky Mountains laid out their T-shirts the night before so they wouldn't forget.

2. Janis, Ashley, and the girl with the Rocky Mountains shirt had their moms help them choose the best shirt for Earth Day.

3. The hummingbird shirt was not worn by Janis.

	hummingbirds	rabbits	world	Rocky Mountains
Janis				
Charlie				
Nicole				
Ashley				

Scrambled Words

Mrs. Harman gave Greg, Vannessa, Brad, and Lauren four letters (I,L,O and S) to arrange into a word. Each came up with a different combination. Their words were *Lois*, *oils*, *silo*, and *soil*. Unscramble these clues to find out who came up with which word.

Clues

1. Greg, and the girl forming *Lois*, and the boy who found *silo*, and Lauren were surprised by their classmates' combinations.

2. Lauren did not come up with the word *soil*.

	Lois	oils	silo	soil
Greg				
Vannessa				
Brad				
Lauren				

Dentist Appointments

Wesley, Claire, Sonja, and Brent all have appointments at Dr. Liebig's dental office on the same day at 2:30, 4:00, 4:30, and 5:00. Use these clues to find out when each is scheduled to see the dentist.

Clues

1. Wesley is scheduled to see Dr. Liebig after Claire and Sonja but before Brent.

2. Sonja's appointment is before Claire's.

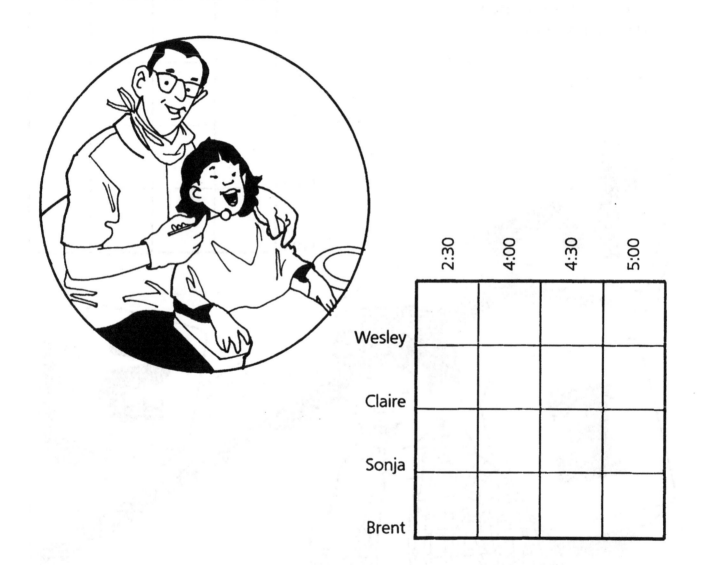

	2:30	4:00	4:30	5:00
Wesley				
Claire				
Sonja				
Brent				

Baby-sitting for Tommy

Marilyn, Kathy, Linda, and Mandy occasionally baby-sit Tommy. The last time each girl took care of him, she brought something special to entertain him. One girl brought a movie, one brought a card game, one came with squirt guns, and another shared her water colors. Sort through the clues to find out which girl brought each item.

Clues

1. Marilyn, Kathy, and the girl with the movie, and the girl with the squirt guns all live in Tommy's neighborhood.

2. Mandy, and the girl with the movie, and the girl with the card game, and the girl with the water colors all attended the same child care course.

3. Marilyn did not have the water colors.

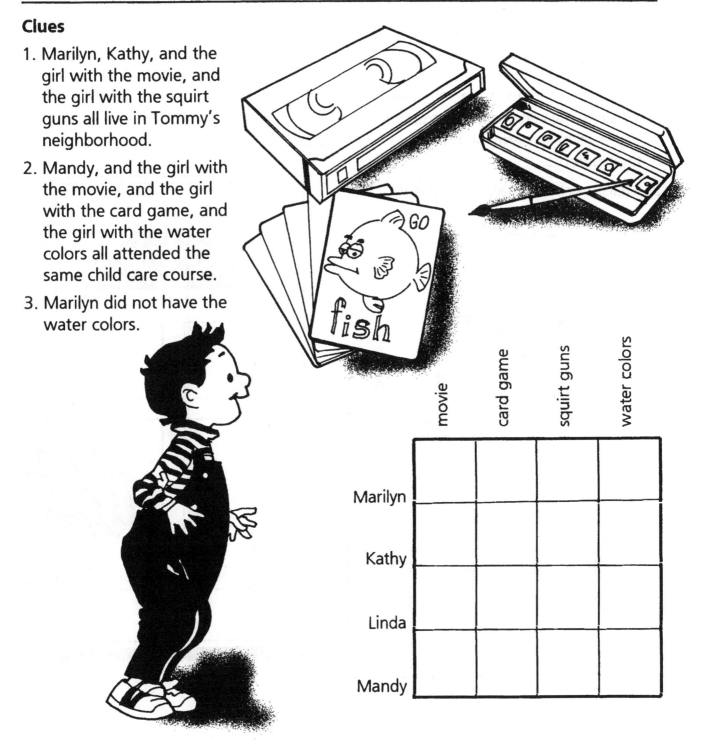

	movie	card game	squirt guns	water colors
Marilyn				
Kathy				
Linda				
Mandy				

May Baskets

Connie, Kyle, Melissa, and Karen have decided to secretly hang May baskets on neighbors' doors to celebrate May Day. Each is contributing something special for the project. They are bringing lilacs, roses, irises, and baskets. Pick your way through these clues to find out who contributed what.

Clues

1. Connie, the boy who will gather roses from the bushes behind his house, the girl cutting irises from her mother's garden, and the girl weaving baskets from construction paper all have permission from their parents.

2. Melissa will not make the baskets.

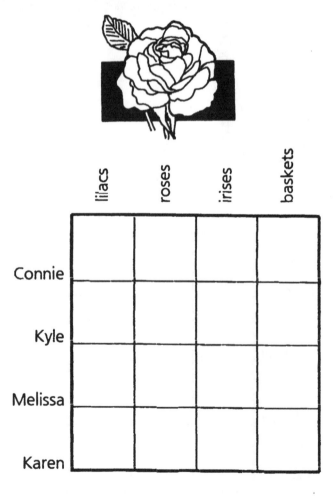

Lawn Service

Chad, Bryan, Mike, and Raul have decided to earn extra money by mowing lawns. Each has chosen to do a different job. The jobs are mowing, weeding, blowing away the grass clippings, and trimming shrubs. Cut through the clues to find out which boy will do each task.

Clues

1. Chad, his brother using the weed whacker, the boy with the blower, and the boy with the electric hedge trimmer all have been taught how to safely use the lawn equipment.

2. Bryan, and the boy with the weed whacker, and the boy with the hedge trimmer, and the boy using the mower borrowed money to buy equipment. Their loans must be paid back before they can make a profit.

3. Raul does not trim shrubs.

	weed	blow away clippings	trim shrubs	mow
Chad				
Bryan				
Mike				
Raul				

Animal Adoption

Students at Claymont School are going to adopt an animal at the St. Louis Zoo. There are four animals they can adopt. They are the black tailed prairie dog, the American alligator, the American black bear, and the bald eagle. Sarah, Nick, Kristi, and Jolene voted for a different animal. Tally these clues to find out who voted for each animal.

Clues

1. Sarah, and the boy who voted for the bald eagle, and the girl supporting the prairie dog, and the girl who wanted to adopt an American alligator all hope their choice will win.

2. Kristi did not vote for the prairie dog.

	prairie dog	alligator	black bear	bald eagle
Sarah				
Nick				
Kristi				
Jolene				

Everyone's a Winner

Several members of the third grade worked at a booth at the carnival. Sean, Laquita, Nathan, and Heather worked together collecting money, handing out balls, giving out prizes, and setting up targets. Use these clues to find out who did what job.

Clues

1. Laquita, and the girl who collected money, and the boy who handed out balls, and the boy who set up targets really enjoyed working in the booth.

2. Nathan did not hand out balls.

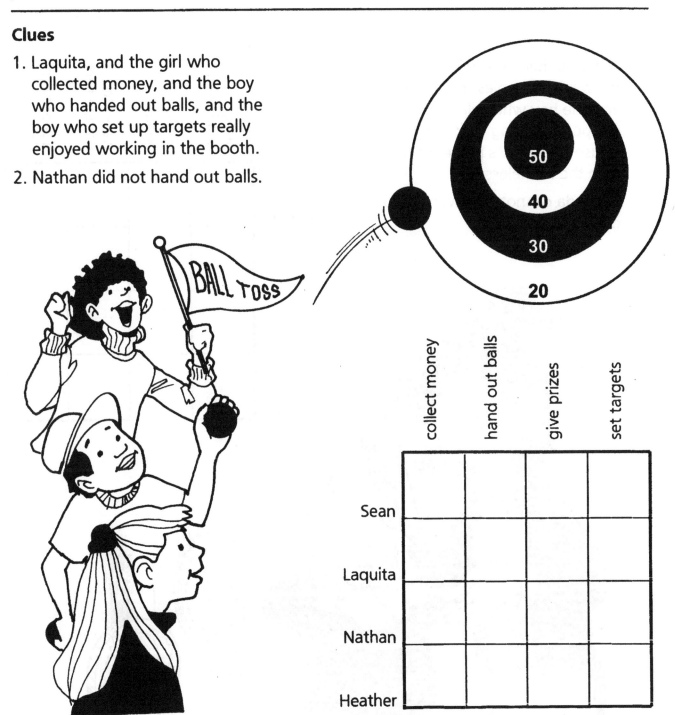

	collect money	hand out balls	give prizes	set targets
Sean				
Laquita				
Nathan				
Heather				

Magic Show

Amanda, Charlie, Leslie, and Stuart are putting on a magic show featuring their best tricks. The tricks are the amazing rising handkerchief, raining pennies, changing water into orange juice, and the disappearing dog. Focus on the hocus-pocus clues to find out which child is performing which trick.

Clues

1. Amanda, her sister who is doing the raining pennies trick, Charlie, and the boy who will do the disappearing dog trick are going to use the show's proceeds to help the homeless.

2. Amanda cannot turn water into orange juice.

	rising handkerchief	raining pennies	water to orange juice	disappearing dog
Amanda				
Charlie				
Leslie				
Stuart				

Time Line

Jason, Steven, Robert, and Rosa are making a time line in Mr. Peterson's class. Each one is drawing a different geological age — Jurassic, Paleozoic, Mississippian, and Pre-Cambrian. Excavate the clues to find out which student is responsible for which geological age.

Clues

1. Jason, and the girl doing the Jurassic Age, and the boy doing the Mississippian Age all talked with the librarian about checking out books to help with the project.

2. The student doing the Pre-Cambrian Age and Jason have decided to work on a diorama to go along with the time line.

3. Robert is not doing the Mississippian Age.

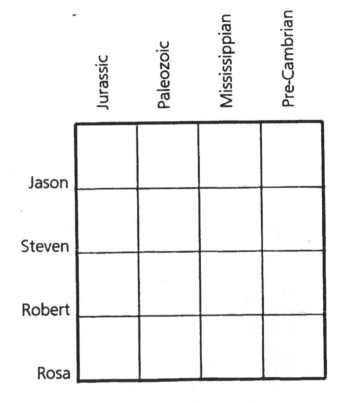

	Jurassic	Paleozoic	Mississippian	Pre-Cambrian
Jason				
Steven				
Robert				
Rosa				

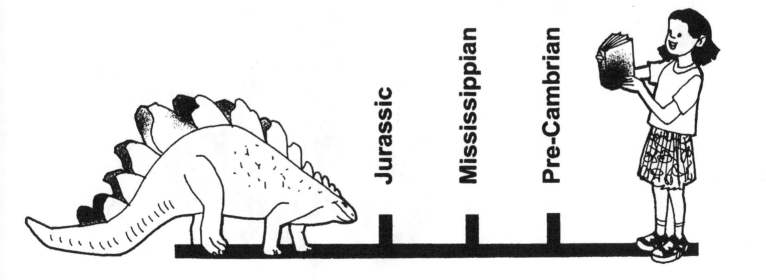

Barber Shop Quartet

Four friends (Tamara, Jarrett, Ben, and Jessica) went to the barber shop together to get their hair cut. Each sat in a different chair with a different barber. The barbers were Teresa, Ralph, Terry, and Andy. Trim away the clues and find out who went to each barber.

Clues

1. Tamara, the boy seeing Ralph, the girl seeing Terry, and Ben go to the same barber each time they get their hair cut.

2. Ben didn't have Andy cut his hair.

	Teresa	Ralph	Terry	Andy
Tamara				
Jarrett				
Ben				
Jessica				

Black History

Wade, Terrell, Drew, and Corey are writing reports on George Washington Carver, Mary McLeod Bethune, Jesse Owens, and Martin Luther King, Jr. for Black History Week. Read the clues to find out who is reporting on each famous person.

Clues

1. Wade, Terrell, and the student writing a report on Mary McLeod Bethune, and the student reporting on Jesse Owens plan to go to the school library and the public library.

2. Drew is friends with Corey and the students researching George Washington Carver and Martin Luther King, Jr.

3. Drew is the only student reporting on a woman.

4. Terrell is not doing a report on Martin Luther King, Jr.

	George Washington Carver	Mary McLeod Bethune	Jesse Owens	Martin Luther King, Jr.
Wade				
Terrell				
Drew				
Corey				

Synonyms

Ms. Washburn asked Crystal, Keisha, Heather, and Meredith to think of synonyms for the word "sew." The words the girls thought of were *baste*, *stitch*, *seam*, and *tack*. Unravel the clues to see which girl thought of which synonym.

Clues

1. Crystal, Heather, and the girl who thought of *baste*, and the girl who said *seam* all love reading.

2. Keisha, and the girl who answered *baste*, and the girl who thought of *tack*, and Crystal are good spellers.

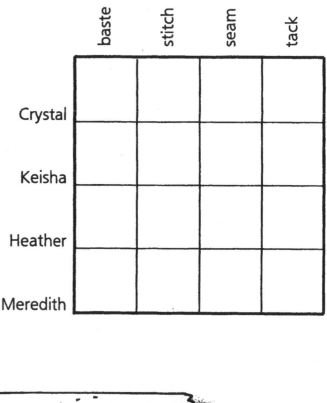

	baste	stitch	seam	tack
Crystal				
Keisha				
Heather				
Meredith				

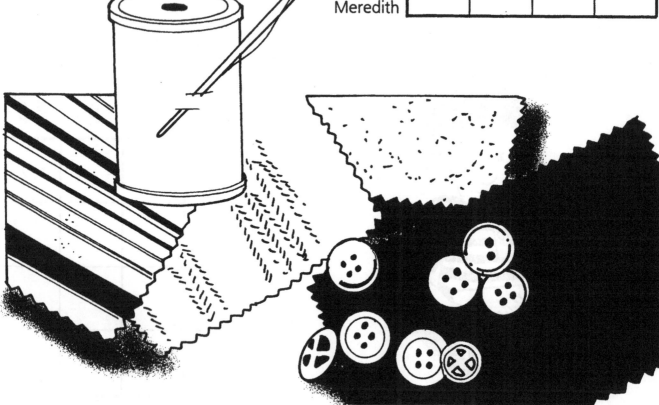

Pool Party

Joshua, Ashley, Kristen and Jeremy were invited to Jack's pool to swim. Each one brought a special item to the pool. They brought diving rings, goggles, a ball, and squirt guns. Dive into these clues and surface with the answer to which friend brought what thing to the pool party.

Clues

1. Joshua, and the girl who brought the ball, and the boy with the diving rings, and the girl with the goggles are special friends of Jack.

2. Kristen did not bring the goggles.

	diving rings	goggles	ball	squirt guns
Joshua				
Ashley				
Kristen				
Jeremy				

Neighborhood Picnic

Brent, Janet, Christopher, and Nina brought barbecued beef, potato salad, deviled eggs, and lemon pie to a neighborhood picnic. Use the clues to find out who brought each dish.

Clues

1. Brent, and the girl who made lemon meringue pie, and the boy who brought barbecued beef are neighbors on Parkrose Court.

2. Janet asked the boy who brought potato salad to walk with her to the picnic.

3. Nina did not bring the deviled eggs.

	barbecued beef	potato salad	deviled eggs	lemon pie
Brent				
Janet				
Christopher				
Nina				

Pizza Connection

Elizabeth, Willy, Margaret, and Seth won pizza certificates for being finalists in the spelling bee. Each student ordered a pizza with a different topping. The toppings were pepperoni, extra cheese, black olives, and sausage. Slice through the clues to find out who ordered which topping.

Clues

1. Willy, the girl who ordered pepperoni, the boy who got black olives, and Elizabeth are all great spellers.

2. Elizabeth did not order sausage.

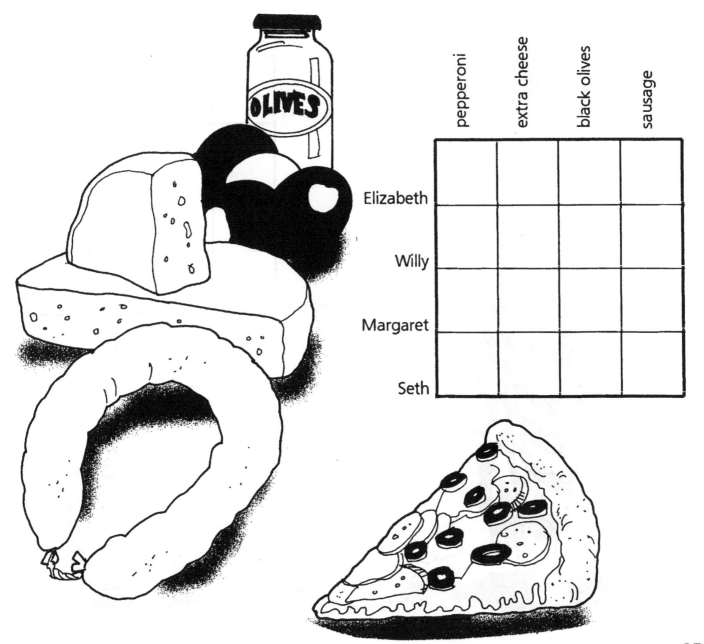

	pepperoni	extra cheese	black olives	sausage
Elizabeth				
Willy				
Margaret				
Seth				

Class Skit

Mr. Ryan's class put on a production of *Aladdin and the Magic Lamp*. Roberta, Maggie, Vannessa, and Michael played the parts of Aladdin, his mother, the evil magician, and the princess. Polish up the clues to find out who played which role.

Clues

1. Roberta, and the girl who played the mother, and Vannessa, and the boy who played Aladdin all helped write the script.
2. The princess was not played by Roberta.

	Aladdin	mother	evil magician	princess
Roberta				
Maggie				
Vannessa				
Michael				

Favorite Video Games

Alex, Miguel, Joanna, and Emily enjoy playing video games. Their favorite games are *Droid Quest, Luna Hunter, Tasheback,* and *Super Astro World*. Scan the clues to find out which game is each person's favorite.

Clues

1. Joanna, her brother who loves *Droid Quest*, her best girl friend who enjoys *Tasheback*, and the boy who favors *Super Astro World* all have time limits that their parents set for playing video games.

2. Alex and Joanna are not related.

	Droid Quest	Luna Hunter	Tasheback	Super Astro World
Alex				
Miguel				
Joanna				
Emily				

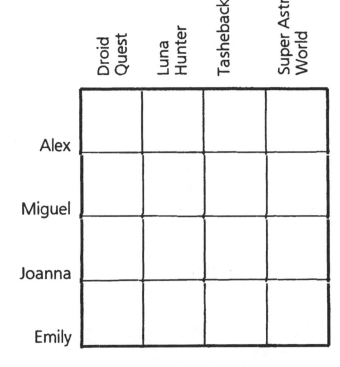

Recycling

Parkway School District began a recycling program in September. During the first semester cardboard, white paper, computer paper and plastic in the amounts of 30 tons, 16 tons, 13 tons, and 5 tons were recycled. Rummage through these clues to discover how much of each thing was recycled.

Clues

1. More white paper was recycled than computer paper, but there was less white paper than cardboard and plastic.

2. Less cardboard was recycled than plastic.

	30 tons	16 tons	13 tons	5 tons
cardboard				
white paper				
computer paper				
plastic				

Butterfly Garden

Mrs. Buck's class is planting a garden to attract colorful butterflies. Linda, Deborah, Josie, and Veronica are supplying the phlox, salvia, milkweed, and petunia plants. Flutter by these clues to discover who is bringing which plants.

Clues

1. Linda, and the girl whose mother thinned her phlox to provide plants to share, and the girl who bought two dozen petunias because her mother didn't have any plants, and Josie are all helping to prepare the soil and plan the garden's design.

2. Deborah, and the girl bringing salvia, and the girl supplying milkweed, and Veronica must take turns watering the garden during summer break.

3. Deborah's mother has no plants to share.

4. Linda lives in town and is unable to collect milkweed.

	phlox	salvia	milkweed	petunias
Linda				
Deborah				
Josie				
Veronica				

Hobby Night

The parent-teacher organization is hosting a family hobby night. Doug, Lee, Sarah, and Rhonda will share their hobbies of stamp collecting, baseball card collecting, bee keeping, and collecting old greeting cards. Sort through the clues to find out which hobby each person enjoys.

Clues

1. Lee and his brother both collect things as a hobby.

2. Sarah inherited her greeting cards from her great aunt.

3. Doug admires his brother's stamps.

	stamp collection	baseball card collection	bee keeping	greeting cards
Doug				
Lee				
Sarah				
Rhonda				

February Mural

Juan, Maria, David, and Sherita are working on a bulletin board for February. Each one is in charge of a different special day. Their days are Groundhog Day, Presidents' Day, Valentine's Day, and the origin of leap year. Leap through the clues to find out who is in charge of sharing information about each occasion.

Clues

1. Juan, Maria and David are working on displays of special February holidays.

2. Juan, the girl making lots of pink and red hearts, and the boy showing a woodchuck are all using paints and construction paper to make their displays.

	Groundhog Day	President's Day	Valentine's Day	leap year
Juan				
Maria				
David				
Sherita				

Snow Day

Eric, Katlin, Brian, and Laura did not have to go to school one day last week. The school was closed because of eight inches of fresh snow. Each person spent the day either sledding, taking part in snow ball fights, earning money by shoveling walks, or building snowmen. Plow through the clues to discover who did each activity.

Clues

1. Eric, the girl who went sledding, the boy who earned money shoveling walks, and Laura all enjoyed the day off.

2. The girl who built snowmen put mufflers and earmuffs on all her creations.

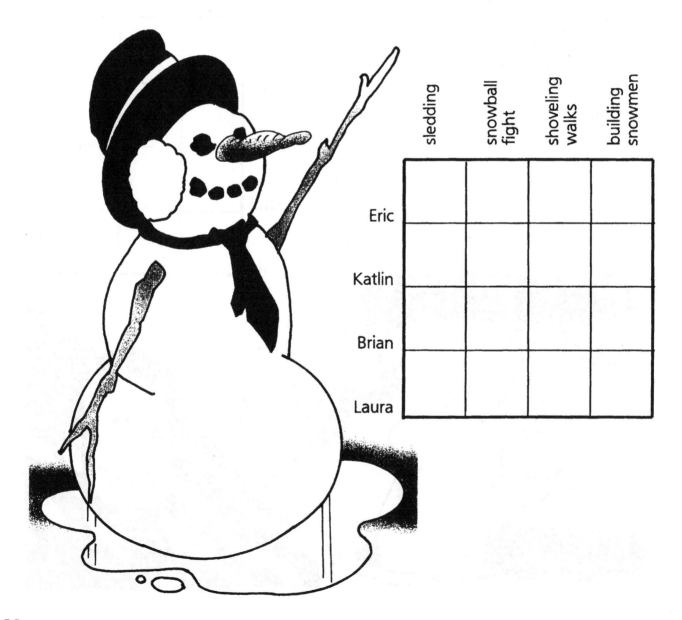

	sledding	snowball fight	shoveling walks	building snowmen
Eric				
Katlin				
Brian				
Laura				

Answers

1. Birthday Bash, pg. 4
Lyle - miniature golf
Ryan - pizza parlor
Troy - roller rink

2. Swimming Lessons, pg. 5
Steve - pennies
Ted - treading water
Andy - diving board

3. After School, pg. 6
Daralynn - puzzles
Kevin - yo-yo
Tara - jump rope

4. Spring Musical, pg. 7
Megan - rose bud
Adam - fisherman
Jeff - baseball player

5. Aluminum Can Drive, pg. 8
Carla - 3 bags
Patrick - 5 bags
Justin - 2 bags

6. Helping In Grandfather's Garden, pg. 9
Tyler - transplanting
Jack - adding compost
Brenda - staking tomatoes

7. Earth Day, pg. 10
Janis - rabbits
Nicole - Rocky Mountains
Charlie - world
Ashley - hummingbirds

8. Scrambled Words, pg. 11
Greg - soil
Vannessa - Lois
Brad - silo
Lauren - oils

9. Dentist Appointments, pg. 12
Wesley - 4:30
Claire - 4:00
Sonja 2:30
Brent 5:00

10. Baby-sitting Tommy, pg. 13
Marilyn - card game
Kathy - water colors
Linda - movie
Mandy - squirt guns

11. May Baskets, pg. 14
Connie - lilacs
Kyle - roses
Melissa - irises
Karen - baskets

12. Lawn Service, pg. 15
Chad - mow
Raul - weed
Bryan - blow away grass clippings
Mike - trim shrubs

13. Animal Adoption, pg. 16
Sarah - black bear
Nick - bald eagle
Kristi - alligator
Jolene - prairie dog

14. Everyone's a Winner, pg. 17
Laquita - give prizes
Nathan - set targets
Heather - collect money
Sean - hand out balls

15. Magic Show, pg. 18
Amanda - rising handkerchief
Charlie - water to orange juice
Leslie - raining pennies
Stuart - disappearing dog

16. Time Line, pg. 19
Jason - Paleozoic
Rosa - Jurassic
Steven - Mississippian
Robert - Pre-Cambrian

17. Barber Shop Quartet, pg. 20
Tamara - Andy
Jarrett - Ralph
Ben - Teresa
Jessica - Terry

18. Black History, pg. 21
Wade - Martin Luther King, Jr.
Terrell - George Washington Carver
Drew - Mary McLeod Bethune
Corey - Jesse Owens

19. Synonyms, pg. 22
Crystal - stitch
Keisha - seam
Heather - tack
Meredith - baste

20. Pool Party, pg. 23
Joshua - squirt guns
Ashley - goggles
Kristen - foam ball
Jeremy - diving rings

21. Neighborhood Picnic, pg. 24
Brent - potato salad
Nina - lemon pie
Janet - deviled eggs
Christopher - barbecued beef

22. Pizza Connection, pg. 25
Elizabeth - extra cheese
Willy - sausage
Margaret - pepperoni
Seth - black olives

23. Class Skit, pg. 26
Roberta - evil magician
Maggie - mother
Vannessa - princess
Michael - Aladdin

24. Favorite Video Games, pg. 27
Alex - Super Astro World
Miguel - Droid Quest
Joanna - Luna Hunter
Emily - Tasheback

25. Recycling, pg. 28
cardboard - 16 tons
white paper - 13 tons
computer paper - 5 tons
plastic - 30 tons

26. Butterfly Garden, pg. 29
Linda - salvia
Deborah - petunia
Josie - milkweed
Veronica - phlox

27. Hobby Night, pg. 30
Doug - baseball cards
Lee - stamp collection
Sarah - greeting cards
Rhonda - bee keeping

28. February Mural, pg.
Juan - President's Day
Maria - Valentine's Day
David - Groundhog Day
Sherita - leap year

29. Snow Day, pg. 32
Eric - snowball fight
Katlin - sledding
Brian - shoveling walks
Laura - building snowmen

Common Core State Standards Alignment Sheet
Logic Safari (Book 1)

All lessons in this book align to the following standards.

Grade Level	Common Core State Standards in ELA-Literacy
Grade 2	RF.2.3 Know and apply grade-level phonics and word analysis skills in decoding words. RF.2.4 Read with sufficient accuracy and fluency to support comprehension.
Grade 3	RF.3.3 Know and apply grade-level phonics and word analysis skills in decoding words. RF.3.4 Read with sufficient accuracy and fluency to support comprehension.

9781593630898

T - #0718 - 101024 - C0 - 276/219/2 - PB - 9781593630898 - Gloss Lamination